Aviation has fascinated and inspired people for generations, and understanding the principles of flight is the first step toward unlocking the mysteries of the skies. Whether you're an aspiring pilot, an aviation enthusiast, or simply curious about how aircraft defy gravity, this book is your guide to grasping the fundamental concepts of flight in just five days.

Learn How Aircraft Flies in 5 Days is designed to make complex aerodynamics accessible to everyone, breaking down the science of flight into clear, manageable lessons that anyone can understand. Over the course of five days, you will explore the key principles that enable an aircraft to take off, stay aloft, and maneuver through the air.

The four fundamental forces acting on an aircraft are thrust, lift, drag, and weight (gravity). We will visit these topics accordingly.

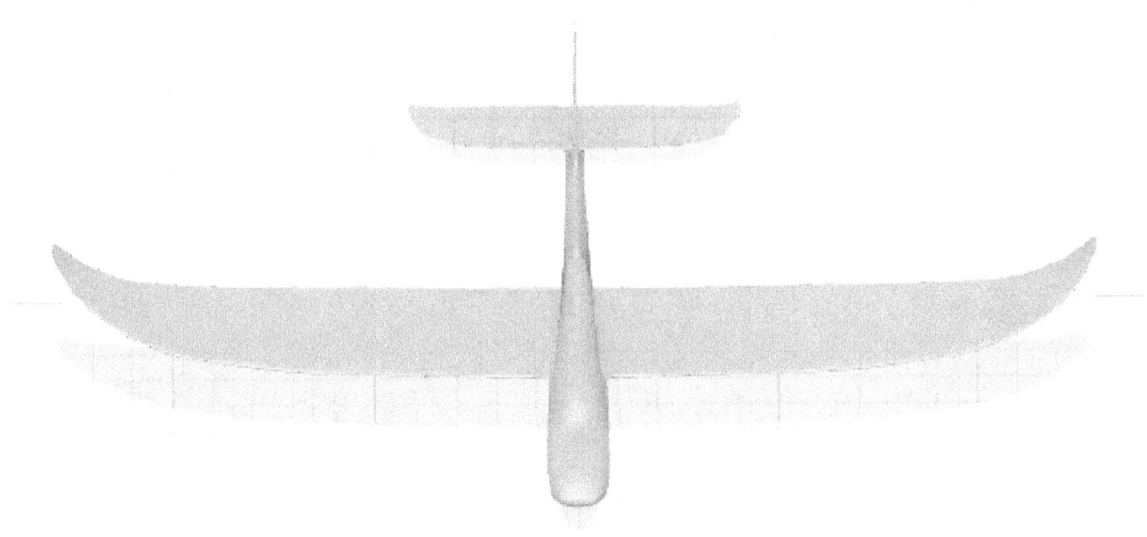

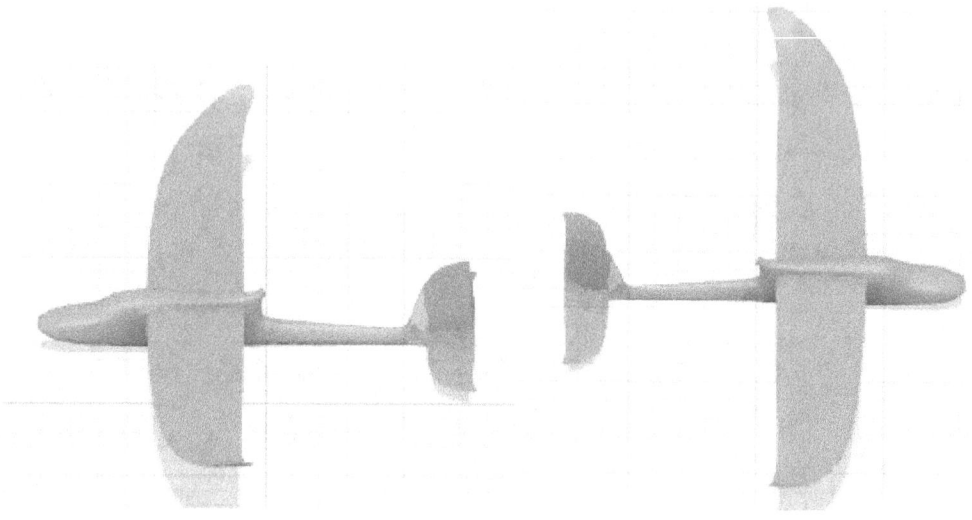

Restrictions on Alteration

You may not modify the Book or create any derivative work of the Book or its accompanying documentation. Derivative works include but are not limited to translations.

Restrictions on Copying

You may not copy any part of the Book unless formal written authorization is obtained from us.

Table of Contents

Day 1 - Basics of Aerodynamics and Lift 4
    Defining Aerodynamics 4
    Defining Airfoil 5
    Defining Bernoulli's Principle 6
    Defining Equal transit time 7
    Lift VS Drag 10
    Pressure Differential 15
    Head wind VS Tail wind 16
    Air speed VS Ground speed 17
Day 2 - Axis, Control Interfaces and Movement 20
    Defining Axis 20
    Defining Airfoil components 23
    Defining Primary Control Surfaces 24

    Defining Primary Control Interfaces 27
    Defining Secondary Control Surfaces 28

Day 3 - Chord Line, CG, CP, AOA and Stall 31
    CG VS CP 32
    Center of Gravity Equals Center of Pressure 33
    Center of Gravity in Front of Center of Pressure 34
    Center of Pressure in Front of Center of Gravity 34
    CG determination 35
    CG and CL 36
    CL and AOA 37
    CP and AOA 37
    Defining Boundary Layer 38
    Flow Separation and Stall 38

Day 4 - More on Lift & Drag, Wingtip vortices, Downwash, Ground Effects and Landing 40
    Aerodynamic center 40
    Relative Wind and AOA VERSUS Critical AoA 40
    Negative AOA and why 41
    Airspeed and AoA 42
    AOI Angle of Incidence 42
    Wing surface area and its relationship with drag & lift 42
    Wingtip vortices, downwash, ground effects, and landing 43
    Is downwash good or bad? 45
    Ground effect during landing 45
    Ground effect during takeoff 46

Day 5 - Stall Speed, Prop Bias, Stabilizers and Flight Stability 47
    Stall speed 47
    Stall and spin 47
    Propeller plane VS Jet stream plane 48
    Prop Wash 49
    Propeller plane left turning tendency 49
    P-Factor (Asymmetric Blade Effect) 50
    Considering all these together 51
    Load factor 52
    Vertical Stabilizer 53
    Horizontal Stabilizer 54

Day 1 – Basics of Aerodynamics and Lift

Defining Aerodynamics

Aerodynamics is the study of how air moves around objects, like airplanes. When an aircraft flies, it interacts with the air in specific ways, and understanding these interactions helps us design planes that can fly smoothly and efficiently.

Here are the basics:

- Lift: Lift is the force that helps an airplane rise off the ground. The wings of the plane are shaped so that air moves faster over the top of the wing and slower underneath. This difference in speed creates a lower pressure on top and higher pressure underneath, lifting the plane into the air (we will talk about this in much greater detail later).
- Drag: Drag is the force that tries to slow the airplane down. It happens because the plane has to push through the air as it moves forward. Designers try to make planes sleek and smooth to reduce drag, allowing them to move faster and use less fuel.
- Thrust: Thrust is the force that pushes the airplane forward. It comes from the engines, which can be jet engines or propellers. The engines generate enough thrust to overcome drag and keep the plane moving forward.
- Weight: Weight is the force that pulls the airplane down toward the Earth. It's caused by gravity. For an airplane to fly, the lift must be greater than the weight.

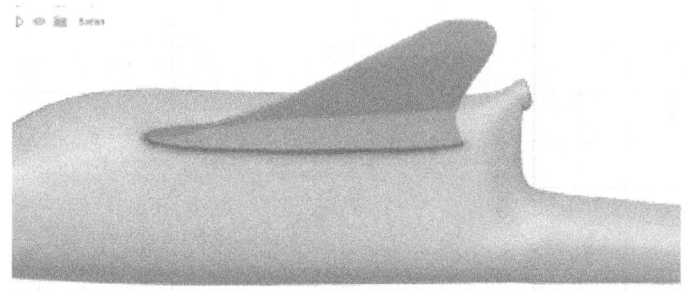

We will talk about lift in this chapter first because this is what is needed to get the plane flying in the air.

Defining Airfoil

The term "airfoil" refers to the cross-sectional shape of the wing. Leading Edge is the front part of the airfoil that first meets the oncoming airflow. It is typically rounded to smooth the flow of air over the wing. **The primary function of an airfoil is to generate lift.**

Symmetrical Airfoils have identical upper and lower surfaces. They generate little or no lift at zero angle of attack and are commonly used in aerobatic aircraft where inverted flight is common. Real world planes do NOT use these. Paper plane does. For a paper plane with a symmetrical airfoil to generate lift, it relies heavily on the angle of attack. The angle of attack is the angle between the wing's chord line (an imaginary line from the leading edge to the trailing edge of the airfoil) and the oncoming airflow (relative wind).

Cambered Airfoils have a curved upper surface and a flatter lower surface. They generate lift more efficiently at lower angles of attack and are typically used in most real world wings.

Modern high speed airplane wings tend to angled backward, a design feature known as sweep or swept-back wings, for several important aerodynamic reasons, particularly related to controlling airflow at high speeds and improving stability. Older and smaller planes often use straight-wing. At high speeds, particularly as an aircraft approaches the speed of sound (known as transonic speeds), the airflow over the wings can reach supersonic speeds even if the aircraft itself is still subsonic. This can create shock waves, which cause a significant increase in drag and potential loss of control.

Swept-back wings help to delay the onset of these shock waves. By angling the wings backward, the airflow is less likely to encounter the wing at a perpendicular angle, effectively reducing the relative speed of the airflow over the wing. This delays the point where the airflow becomes supersonic, allowing the aircraft to fly faster without encountering the detrimental effects of shock waves. Swept-back wings also contribute to a smoother airflow over the wings, which reduces drag. The angle helps to distribute the pressure more evenly along the wing, reducing the overall drag and improving fuel efficiency. This is particularly important for high-speed commercial jets and military aircraft.

Defining Bernoulli's Principle

Bernoulli's Principle is a fundamental concept in fluid dynamics that plays a crucial role in understanding how lift is generated on an aircraft wing. The principle states that in a fluid (which includes both liquids and gases), an increase in the speed of the fluid results in a decrease in its pressure. This relationship between velocity and pressure is essential for explaining how aircraft wings generate lift.

Aircraft wings are designed with a special shape called an airfoil, which is typically curved on the top and flatter on the bottom. This shape forces the air traveling over the top of the wing to move faster than the air traveling underneath it.

According to Bernoulli's Principle, the faster-moving air over the top of the wing generates lower pressure compared to the slower-moving air under the wing.

The difference in pressure between the top and bottom of the wing creates a net upward force called lift. The higher pressure on the underside of the wing pushes the wing upwards, while the lower pressure on the top of the wing pulls it upwards. This combination of forces allows the aircraft to rise and stay in the air.

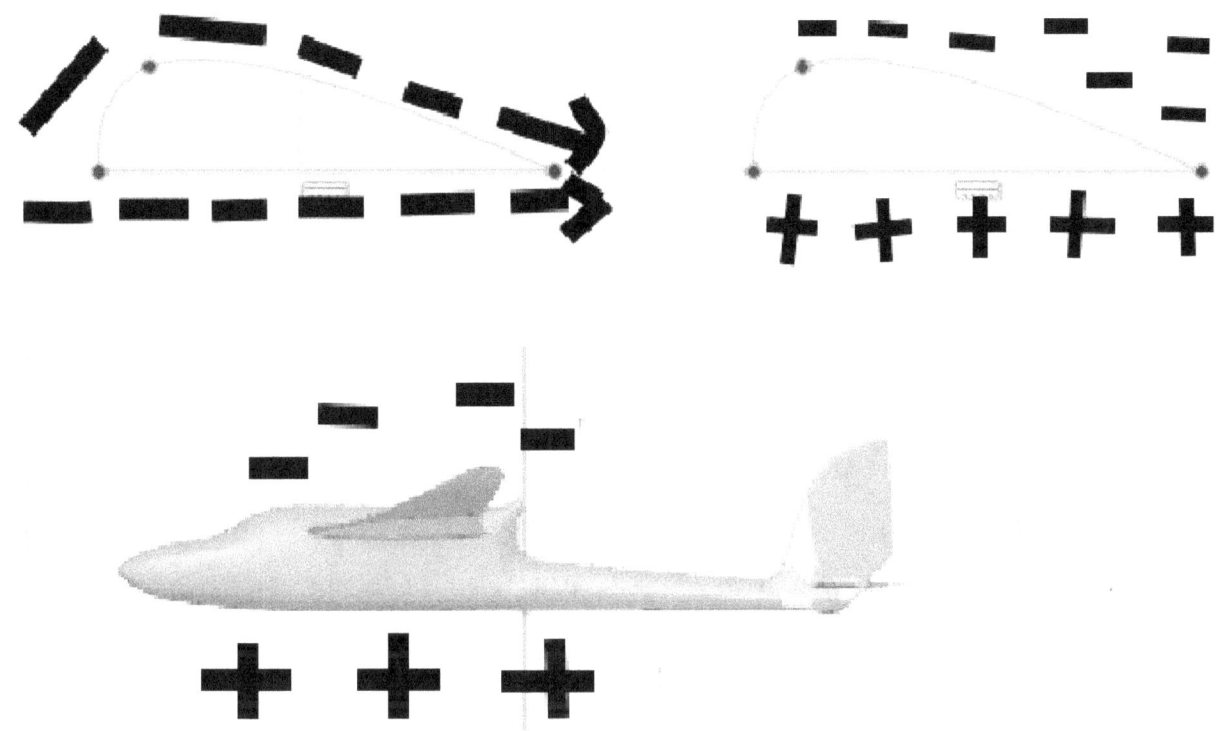

Do note that lift is not solely generated by Bernoulli's Principle, as factors like the angle of attack (the angle at which the wing meets the oncoming air) and Newton's Third Law of Motion (for every action, there is an equal and opposite reaction) also contribute. However, Bernoulli's Principle is a key component in explaining why and how the pressure differential is created.

Defining Equal transit time

The "Equal Transit Time" theory is a common, but incorrect explanation often associated with Bernoulli's Principle when describing how lift is generated on an aircraft wing. It suggests that air molecules split at the leading edge of an aircraft wing (airfoil) and must reunite at the trailing edge at the same time. Because the top surface of the wing is curved (and hence longer) compared to the flat bottom surface, this theory assumes that the air traveling over the top must move faster than the air traveling underneath to meet up at the same point at the trailing edge.

According to Bernoulli's Principle, if the air over the top surface of the wing moves faster, it would indeed result in lower pressure above the wing. The slower air underneath would result in higher pressure, creating a pressure differential that generates lift. HOWEVER, there is no physical principle or law that requires the air molecules to meet up again at the same time at the trailing edge. In reality, the air over the top of the wing moves much faster than the air beneath it, and the two streams do not necessarily meet at the same point simultaneously.

Simply put, the lift is generated due to the shape of the wing and the angle of attack, causing the air over the top to accelerate more than it would if equal transit time were true. This acceleration leads to a significant pressure drop over the top surface, creating lift according to Bernoulli's Principle, but not because of equal transit time.

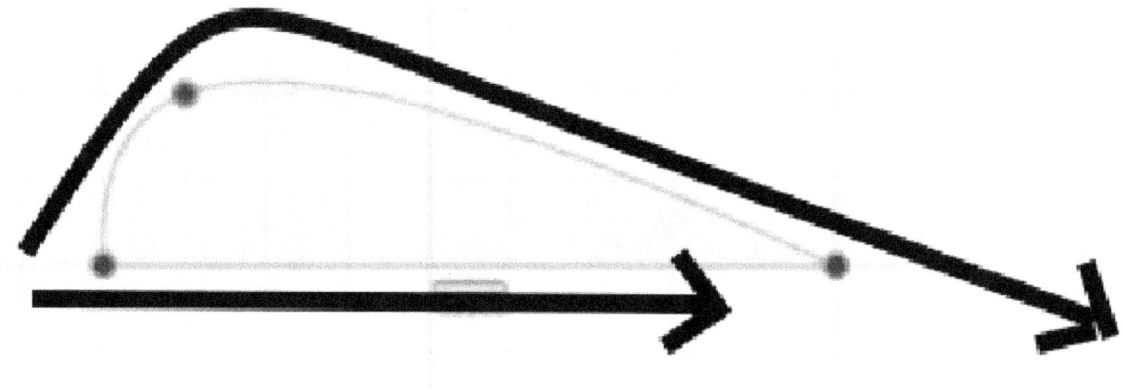

Wind tunnel experiments and fluid dynamics studies have shown that the air over the top of the wing reaches the trailing edge faster than the air traveling underneath, debunking the Equal Transit Time theory.

Defining Newton's Third Law

Newton's Third Law of Motion is highly relevant to understanding how lift is generated on an aircraft. This law states that "for every action, there is an equal and opposite reaction." In the context of lift, this principle plays a crucial role alongside Bernoulli's Principle.

As an aircraft moves forward, its wings (airfoils) interact with the oncoming air. The shape and angle of the wing cause the air to be deflected downward. This deflection is most noticeable under the wing, where the air is pushed downwards as the wing moves through it. According to Newton's Third Law, if the wing exerts a downward force on the air (the action), the air must exert an equal and opposite force on the wing (the reaction). This upward force exerted by the air on the wing is what we recognize as lift.

As the wing pushes air downward, the reaction to this push is the lift that supports the weight of the aircraft. The air that is deflected downward by the wings is known as downwash. This downwash is directly related to the lift force. The more air that is pushed downward, the greater the lift generated by the reaction force.

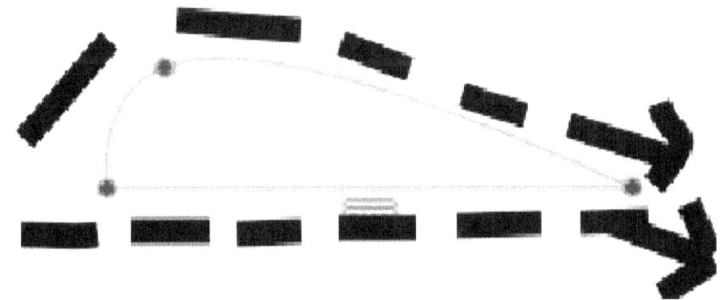

While Bernoulli's Principle explains the pressure differential that contributes to lift, Newton's Third Law explains the lift in terms of momentum. The combination of airflow being deflected downward and a pressure difference helps to generate lift.

Lift VS Drag

Lift and drag are two essential aerodynamic forces that play crucial roles in the flight of an aircraft, alongside weight and thrust. Understanding the interaction between lift and drag helps explain how an aircraft flies, maintains altitude, and maneuvers.

Lift is about up and down "movement" of the plane.

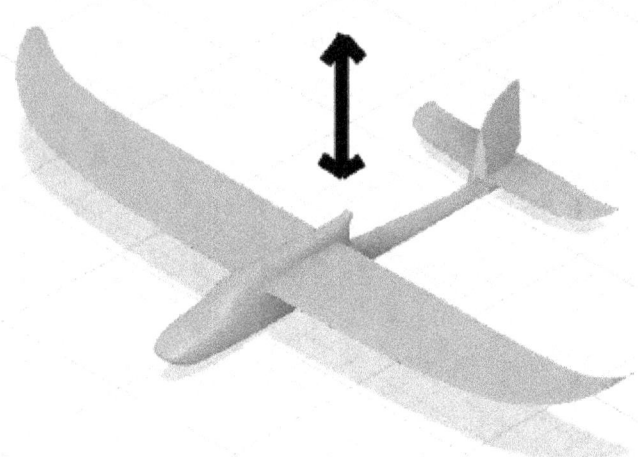

Lift is the force that acts perpendicular to the direction of the aircraft's motion through the air, allowing it to rise off the ground and stay aloft.

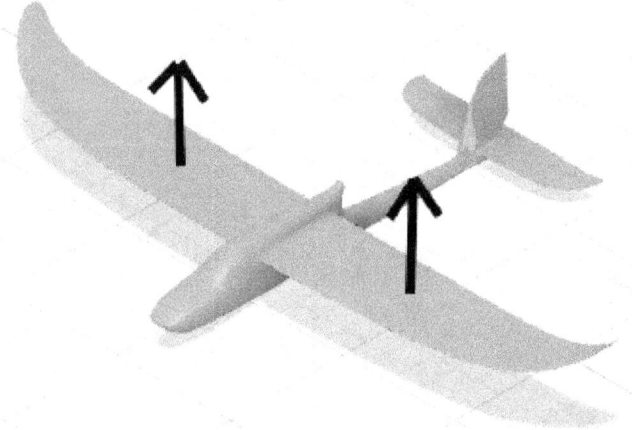

This force is primarily generated by the wings, or airfoils, of the aircraft. As the aircraft moves forward, air flows over and under the wings, creating a pressure difference between the upper and lower surfaces. The shape of the wings causes the air pressure on the top surface to be lower than that on the bottom surface, generating an upward force. This process is explained by a combination of Bernoulli's Principle, which accounts for the pressure difference, and Newton's Third Law, which describes the lift as a reaction to the downward deflection of air by the wings.

Drag, on the other hand, is the force that opposes the aircraft's motion through the air, acting parallel and opposite to the direction of its movement:

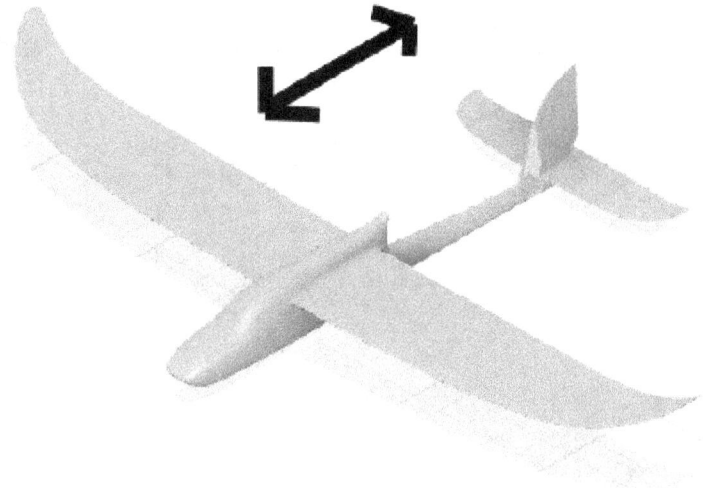

Drag will slow you down:

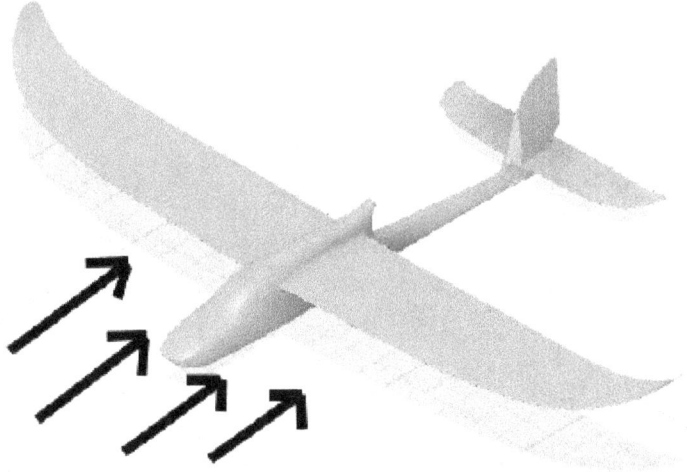

Drag resists the aircraft's forward motion and is divided into two main types: parasite drag and induced drag. Parasite drag arises from factors like the shape of the aircraft, the friction of air flowing over its surfaces, and the interaction of airflow between different parts of the aircraft, such as where the wings meet the fuselage. Induced drag, a byproduct of lift, is caused by the vortices that form at the wingtips as the wings generate lift. These vortices create a pressure differential that contributes to drag. Induced drag is more prominent at lower speeds and higher angles of attack, whereas parasite drag increases with speed.

The relationship between lift and drag is a balancing act. As the angle of attack (we will talk about AOA later) increases to generate more lift, induced drag also increases. This interplay requires careful management in aircraft design and operation to ensure efficient flight. The lift-to-drag ratio, which measures the amount of lift generated per unit of drag, is a key indicator of an aircraft's aerodynamic efficiency. A higher lift-to-drag ratio means the aircraft can achieve more lift with less drag, leading to better fuel efficiency and longer range.

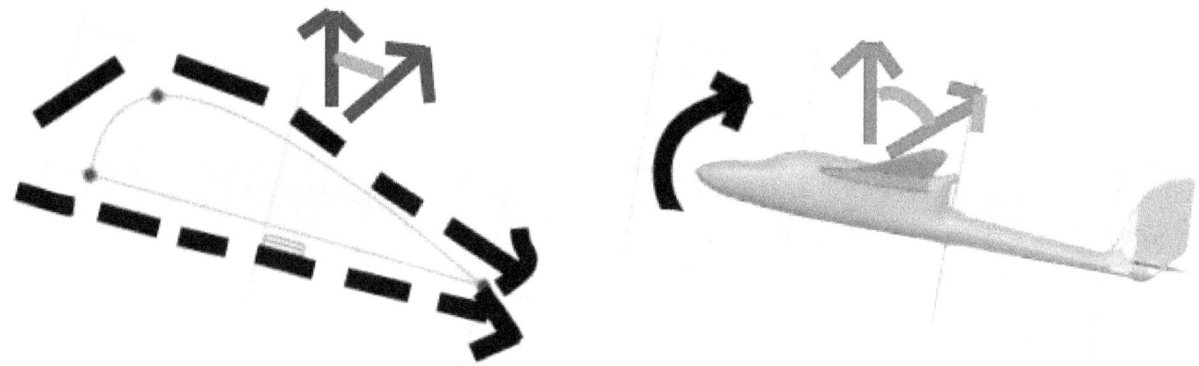

Minimizing parasite drag is crucial for improving an aircraft's aerodynamic efficiency and overall performance. Parasite drag consists of three main components: form drag, skin friction drag, and interference drag. Each of these can be reduced through careful design and maintenance practices.

- The shape of the aircraft significantly affects form drag, which is the resistance caused by the shape of the aircraft as it moves through the air. Streamlining the design to ensure smooth, aerodynamic shapes helps air to flow more easily around the aircraft, reducing form drag. This includes designing fuselages, wing edges, and

other components with gentle curves and tapered shapes that minimize turbulence.
- Skin friction drag arises from the friction between the aircraft's surface and the air. To minimize this, the surface of the aircraft should be as smooth as possible. This can be achieved by using high-quality, smooth materials for the aircraft's skin, regularly cleaning the surface to remove debris or contaminants, and applying special coatings that reduce friction, such as laminar flow control coatings.
- Interference drag occurs where different parts of the aircraft meet, such as where the wings attach to the fuselage or where landing gear is mounted. These intersections can cause airflow disturbances that increase drag. To minimize interference drag, the junctions of these components should be carefully designed and faired (smoothed) to allow air to flow more smoothly. Fillets, which are smooth curves connecting different surfaces, can be used to ease the transition between components and reduce drag.
- Landing gear can create significant drag when extended. Using retractable landing gear that tucks into the aircraft's body during flight reduces drag by eliminating the protruding parts that disrupt airflow.

Induced drag is a by-product of lift (therefore unavoidable) and is associated with the formation of wingtip vortices as the air pressure below the wing interacts with the lower pressure above it. We will talk about this in greater detail later in this book. For now you must know that air not only move from front to back..... it can move across the wingspan and from bottom to top as well (from high pressure area to low pressure area).

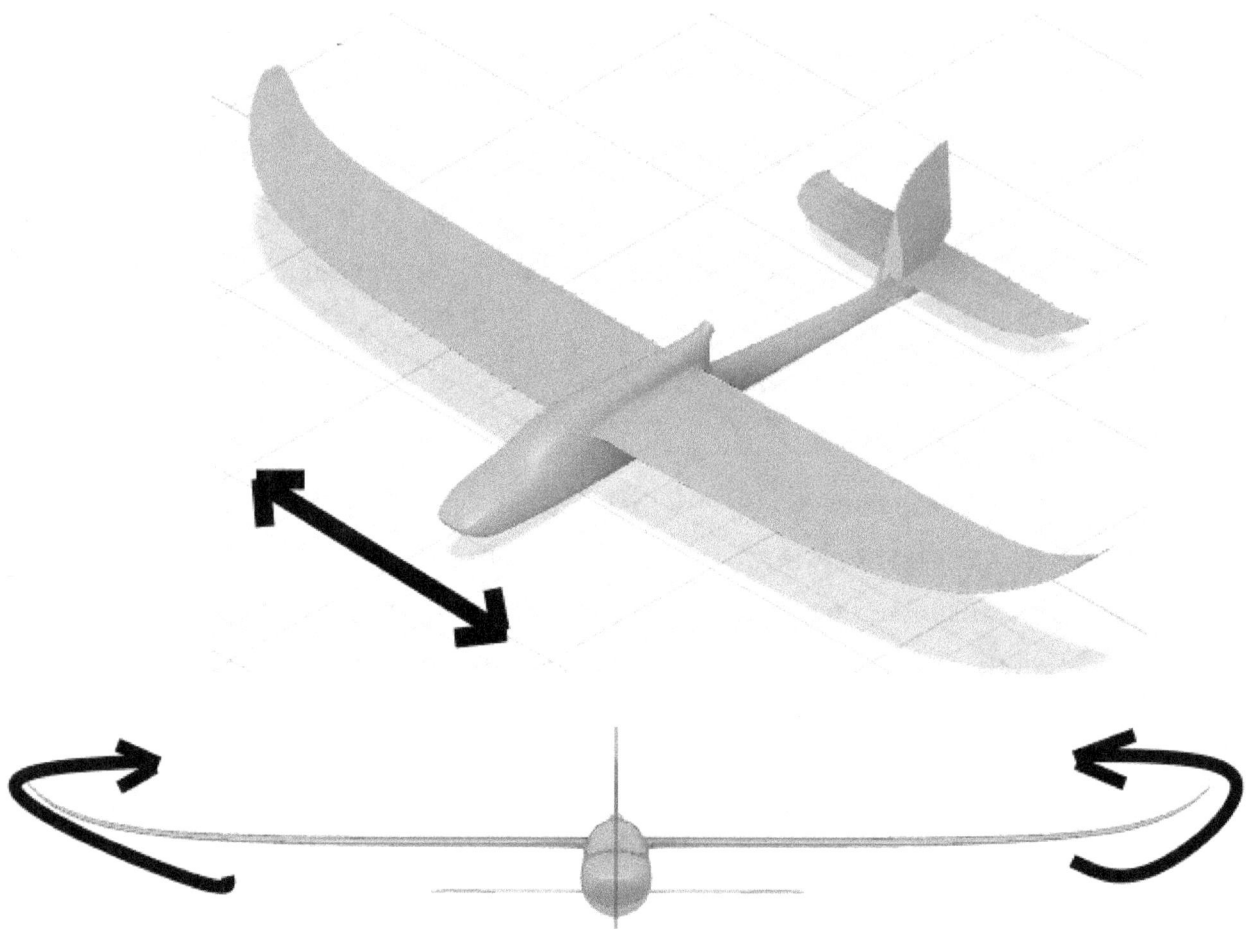

Winglets are vertical or angled extensions at the wingtips that help reduce the formation of wingtip vortices. By limiting the spanwise flow of air from the high-pressure region under the wing to the low-pressure region above the wing, winglets reduce the strength of the vortices, thereby decreasing induced drag. Winglets can also increase the effective wingspan without requiring a longer wing, which helps in reducing drag. On the other hand, a longer wingspan (longer wings that is) increases the aspect ratio (the ratio of the wingspan to the wing chord), which decreases the strength of wingtip vortices. A higher aspect ratio wing generates the same amount of lift with less induced drag. However, longer wings can be more challenging to design and manage structurally, especially for large or high-speed aircraft.

Pressure Differential

The movement of air from high pressure to low pressure is a fundamental concept in fluid dynamics and is driven by the principles of physics, particularly the behavior of gases.

Pressure is essentially the force exerted by air molecules colliding with the surfaces around them. In any given space, when air is compressed or concentrated, it results in a region of high pressure where there are more air molecules in a given volume. Conversely, a region of low pressure has fewer air molecules in the same volume, resulting in less force exerted by these molecules.

Why Air Moves from High to Low Pressure:

Air, like all fluids, naturally moves from areas of higher pressure to areas of lower pressure to reach equilibrium. This movement occurs because of the inherent tendency of molecules to spread out and evenly distribute themselves when free to move. Simply put, because the molecules in the high-pressure region are more crowded, they tend to spread out toward the lower-pressure region where there is more space. This movement is driven by the natural tendency to reduce the concentration of molecules and balance the pressure across the space.

Wind is a result of air moving from high-pressure areas to low-pressure areas in the atmosphere. For example, when there is a high-pressure system and a low-pressure system nearby, air will move from the high-pressure zone to the low-pressure zone, creating wind. Relative airflow refers to the direction and speed of the air moving relative to an object, such as an aircraft, that is moving through the air. It is a key concept in aerodynamics, as it influences the forces acting on the aircraft, including lift, drag, and the behavior of the control surfaces. This direction is always opposite to the direction of the aircraft's flight path. For example, if an aircraft is flying forward, the relative airflow is coming directly toward the nose of the aircraft.

The speed of the relative airflow is essentially the same as the speed of the aircraft through the air, but in the opposite direction. It can change depending on the aircraft's velocity, wind conditions, and altitude. In other words, it can affect how the air interacts with the aircraft's wings, fuselage, and control surfaces. Its relationship with the wing is what generates lift, allowing the aircraft to fly. Similarly, drag is also dependent on the characteristics of the relative airflow.

Head wind VS Tail wind

Headwinds and tailwinds refer to the direction of the wind relative to an aircraft's movement, and they have different impacts on flight.

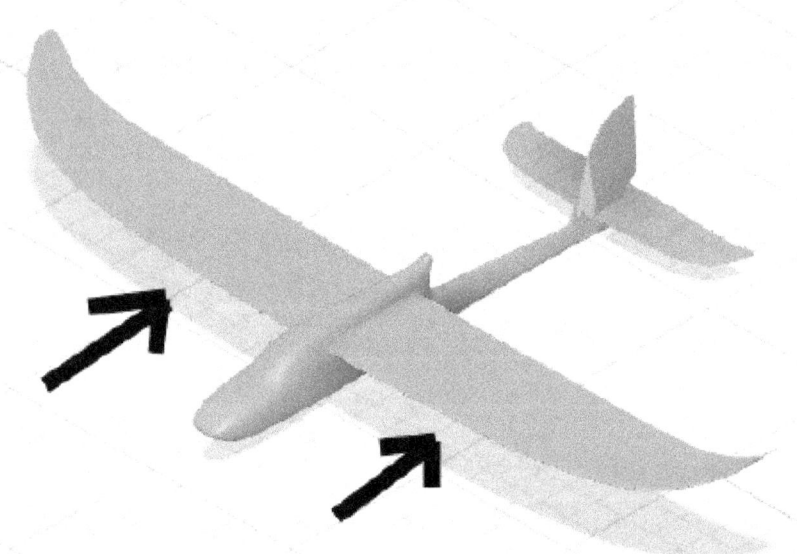

A headwind is a wind that blows directly against the direction in which the aircraft is flying. For example, if a plane is heading north and the wind is coming from the north, the aircraft is facing a headwind. This wind slows the plane's ground speed, which is the speed relative to the ground, because the wind is opposing its motion. Although the aircraft's airspeed, which is the speed through the air, stays the same, the reduced ground speed means the flight could take longer and consume more fuel. However, headwinds are actually beneficial during takeoff and landing. They increase the lift generated by the wings at a given ground speed, which allows the aircraft to take off or land using a shorter distance.

On the other hand, a tailwind is a wind that blows in the same direction as the aircraft's flight. If a plane is flying north and the wind is coming from the south, it's experiencing a tailwind. This wind increases the aircraft's ground speed because it pushes the plane forward. As a result, flights with a tailwind can be faster and more fuel-efficient since the increased ground speed shortens the flight time. However, tailwinds can be challenging during takeoff and landing because they reduce the effective lift at a given ground speed, which requires the aircraft to use longer runways for these operations.

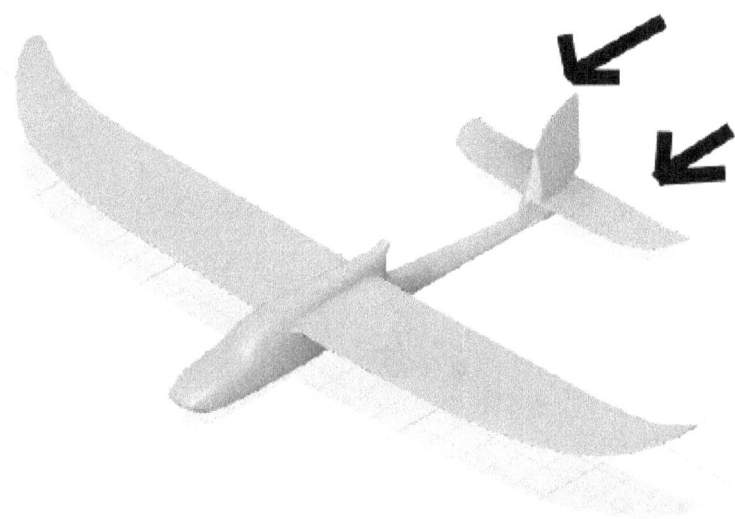

Air speed VS Ground speed

Airspeed is the speed of an aircraft relative to the air around it. It measures how fast the plane is moving through the air, regardless of how fast it is moving over the ground. Airspeed is crucial for flight performance because it directly affects the aircraft's ability to generate lift, which is necessary for flight. The airplane's instruments measure airspeed using a device called a pitot tube, which detects the pressure of the air as the plane moves through it.

Airspeed is important for maintaining safe flight conditions. For instance, an aircraft must reach a certain airspeed to generate enough lift for takeoff. Similarly, the pilot monitors airspeed during flight to avoid stalling, which happens when the plane's airspeed drops too low to generate sufficient lift.

Ground speed is the speed of the aircraft relative to the ground. It is the actual speed at which the airplane is moving over the Earth's surface. Ground speed is affected by the wind: a tailwind increases ground speed, while a headwind decreases it. Unlike airspeed, ground speed is not directly related to the aircraft's ability to stay in the air, but it is important for navigation and determining how long a flight will take. For example, if an aircraft has an airspeed of 300 knots and is flying with a 50-knot tailwind, its ground speed would be 350 knots. Conversely, if the same aircraft is flying into a 50-knot headwind, its ground speed would be reduced to 250 knots.

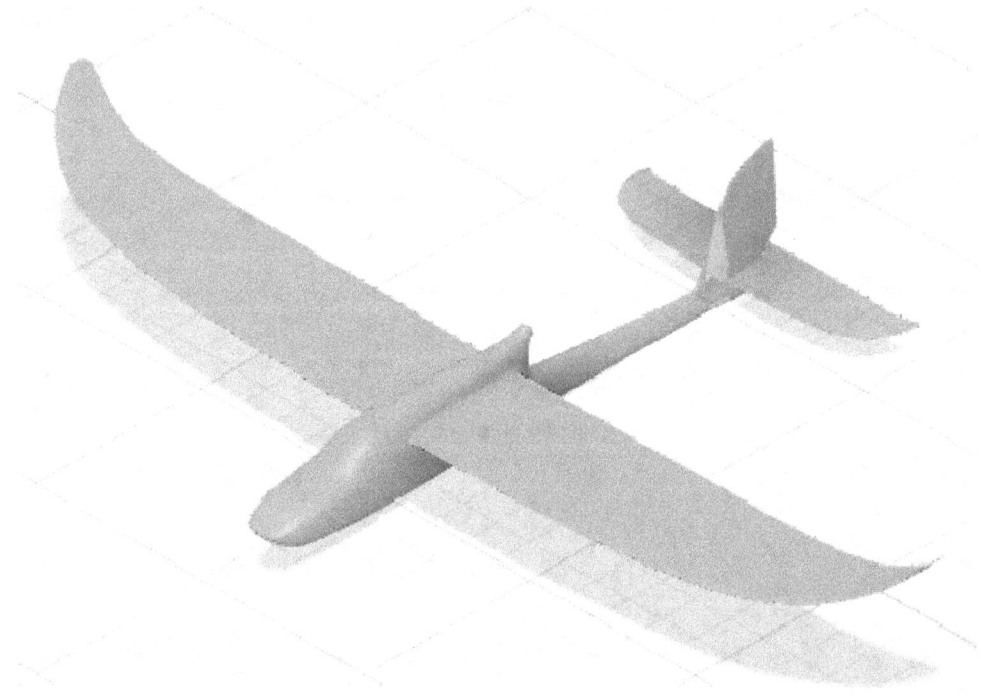

Thrust is a mechanical force generated by the engines of an aircraft to push or pull it forward. It is produced by accelerating a mass of air or exhaust gases backward, according to Newton's Third Law of Motion, which states that for every action, there is an equal and opposite reaction. The reaction to this accelerated mass of air or gases generates the forward-moving thrust.

Thrust can be broken down into components in relation to aircraft axes:

- Horizontal Component: The part of thrust that acts parallel to the aircraft's longitudinal axis (the axis running from the nose to the tail). This component is primarily responsible for overcoming drag and accelerating the aircraft forward.
- Vertical Component: In certain flight conditions (like during climb or descent), a component of thrust may act vertically, either contributing to lift or opposing weight. This is particularly relevant in aircraft with vectored thrust or during steep climbs.

Day 2- Axis, Control Interfaces and Movement

In aviation, an aircraft's movement is defined by three primary axes, which represent different types of motion. These axes are imaginary lines that pass through the aircraft's center of gravity CG, helping to describe its orientation and behavior during flight. We will talk about CG later in this book, for now let's have a quick summary here:

- Longitudinal Axis (Roll Axis): Controls rolling motion, managed by ailerons.
- Lateral Axis (Pitch Axis): Controls pitching motion, managed by the elevator.
- Vertical Axis (Yaw Axis): Controls yawing motion, managed by the rudder.

Defining Axis

The longitudinal axis runs lengthwise from the nose to the tail of the aircraft. Movement around this axis is known as roll. When an aircraft rolls, one wing tilts upward while the other tilts downward, causing the plane to lean to one side. This rolling motion is controlled by the ailerons, which are located on the wings. By adjusting the ailerons, the pilot can roll the aircraft left or right.

Rolling - left or right tilting of the wing:

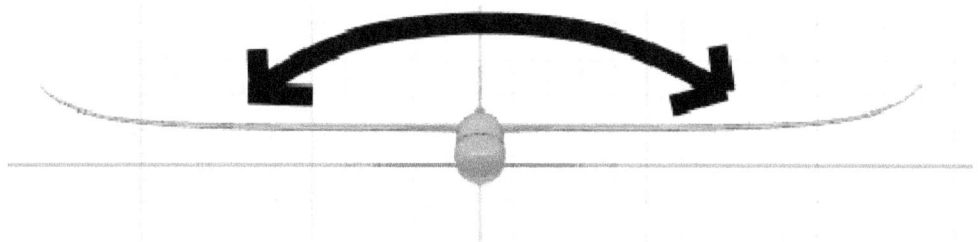

The lateral axis extends from one wingtip to the other, running perpendicular to the longitudinal axis. Movement around this axis is referred to as pitch. Pitching the aircraft causes the nose to move up or down, which affects whether the plane ascends or descends. The elevator, found on the horizontal stabilizer at the tail, controls pitch. The pilot can adjust the elevator to make the aircraft's nose pitch upward or downward, depending on the desired flight path.

Pitch - nose up or down:

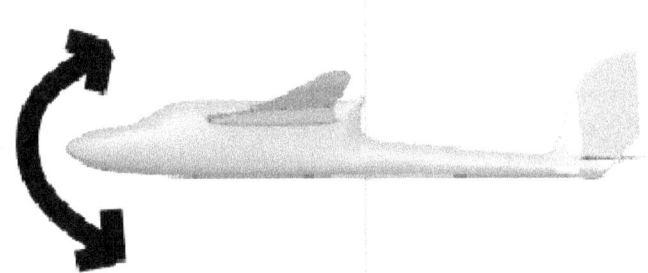

Finally, the vertical axis runs vertically through the aircraft, from top to bottom, intersecting the center of gravity. Movement around this axis is called yaw. Yawing changes the direction the aircraft's nose is pointing without tilting the wings, essentially swiveling the plane left or right. This motion is managed by the rudder, located on the vertical stabilizer (the fin) at the tail. The pilot can move the rudder left or right to adjust the aircraft's yaw and change its direction.

Yaw - change direction without wing tilting:

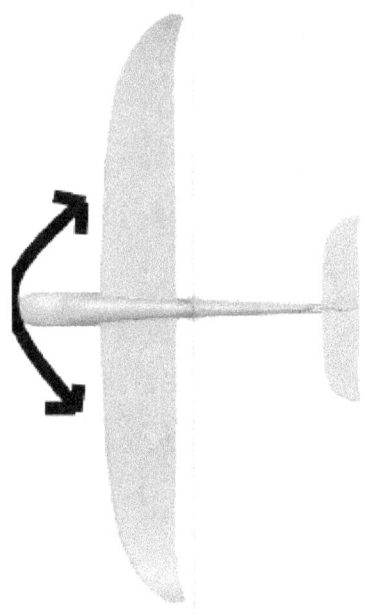

Banking is the term used to describe the tilting of an aircraft to one side during a turn. When a plane banks, one wing dips lower than the other, creating an angled position relative to the horizon. This maneuver is essential for changing direction smoothly in flight.

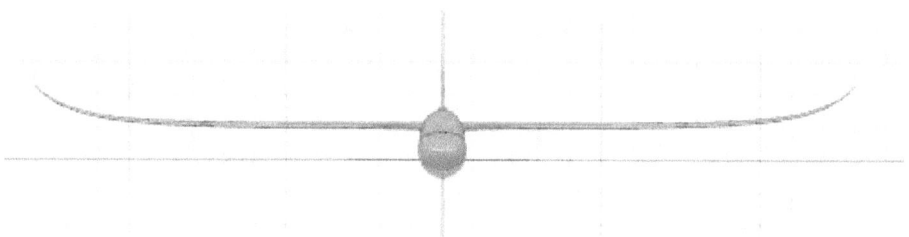

In fact, banking is the result of rolling the aircraft.

Roll refers to the rotation of the aircraft around its longitudinal axis, which runs from the nose to the tail. When a pilot rolls the aircraft, they tilt the wings either up or down, causing the plane to lean to one side or the other. Banking is the specific outcome of rolling the aircraft to initiate a turn.

Here's how the process works in greater details:
- When a pilot wants to turn the aircraft to the left or right, they use the ailerons—control surfaces located on the trailing edge of each wing. By moving the control stick or yoke to one side, the ailerons on one wing move up while those on the opposite wing move down. This action changes the lift generated by each wing: the wing with the raised aileron produces less lift, causing it to drop, while the other wing with the lowered aileron produces more lift, causing it to rise. As a result, the aircraft tilts, or "banks," in the direction of the desired turn.
- During a bank, the lift force that normally acts directly upward is now divided into two components: one that continues to support the aircraft against gravity (vertical component) and another that causes the aircraft to turn (horizontal component). The greater the bank angle, the more significant the turning force, allowing for a tighter turn.
- Banking is crucial for coordinated turns because it allows the aircraft to change

direction smoothly without relying solely on the rudder, which could result in a less stable and less comfortable turn. Proper banking ensures that passengers feel minimal side-to-side forces, and the aircraft maintains a balanced and controlled turn.

Defining Airfoil components

As previously said, the term "airfoil" refers to the cross-sectional shape of the wing. Leading Edge is the front part of the airfoil that first meets the oncoming airflow. It is typically rounded to smooth the flow of air over the wing.

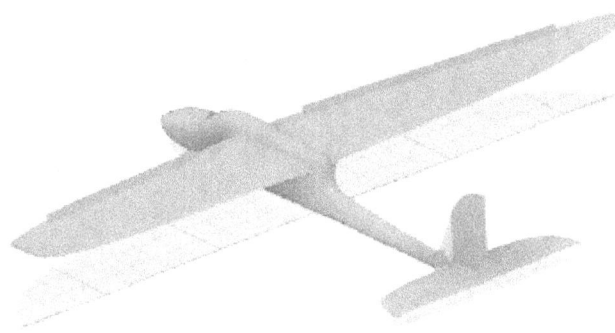

The trailing edge is the rear part of the airfoil where the airflow from the upper and lower surfaces of the wing meet and exit. This edge is usually sharper than the leading edge to help manage airflow separation.

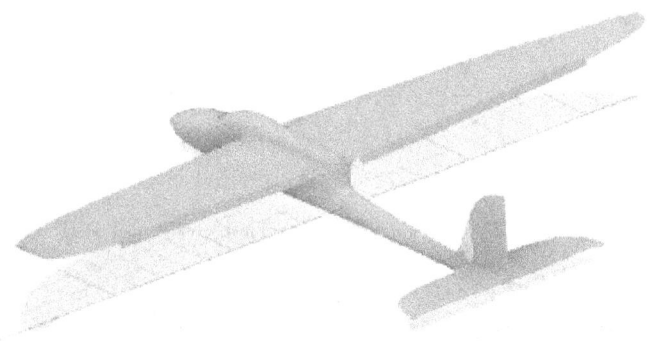

The chord line is an imaginary straight line that connects the leading edge to the trailing edge of the airfoil. The length of this line is known as the chord.

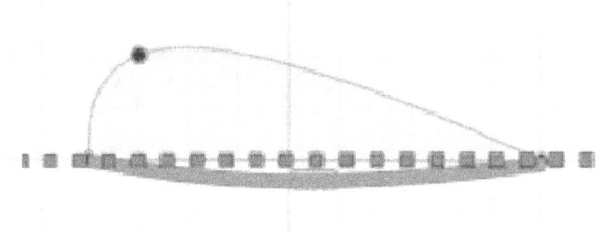

Camber refers to the curvature of the airfoil.

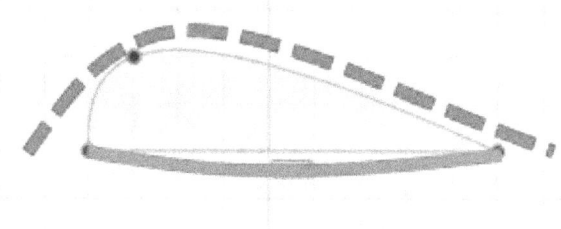

An airfoil can be symmetrical (having no camber, with identical upper and lower surfaces) or asymmetrical (with a curved upper surface and a flatter lower surface). Cambered airfoils are more common in aircraft wings because they generate more lift. The thickness of an airfoil is the distance between the upper and lower surfaces. This varies along the length of the airfoil and influences its aerodynamic properties.

Defining Primary Control Surfaces

<u>A quick summary here:</u>
- Ailerons control roll, allowing the aircraft to bank left or right.
- The elevator controls pitch, allowing the aircraft to climb or descend.
- The rudder controls yaw, allowing the aircraft to change direction left or right.

Ailerons are the control surfaces located on the outer trailing edges of each wing.

They are used to control the aircraft's roll (rotation around the longitudinal axis, which runs from the nose to the tail). They work in pairs, moving in opposite directions. When the pilot turns the control yoke or stick to the right, the right aileron deflects upward while the left aileron deflects downward. This action decreases lift on the right wing and increases lift on the left wing, causing the aircraft to roll to the right. Conversely, turning the yoke or stick to the left causes the aircraft to roll to the left. The roll movement allows the aircraft to bank into a turn. Properly coordinated turns require the use of ailerons along with the rudder to maintain smooth and balanced flight.

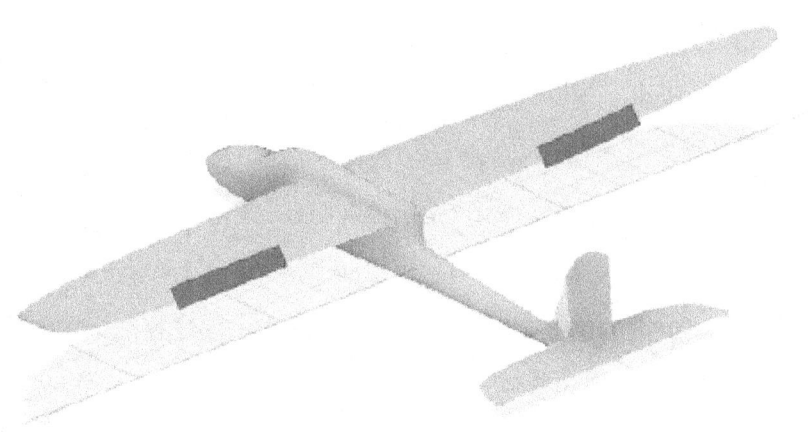

The elevator is the control surface located on the trailing edge of the horizontal stabilizer at the tail of the aircraft. It controls the aircraft's pitch (rotation around the lateral axis, which runs from wingtip to wingtip). It moves up or down in response to the pilot pulling or pushing on the control yoke or stick. Pulling back on the yoke raises the elevator, which increases the downward force on the tail, causing the nose of the aircraft to pitch up. This results in the aircraft climbing. Pushing the yoke forward lowers the elevator, reducing the downward force on the tail and causing the nose to pitch down, which makes the aircraft descend.

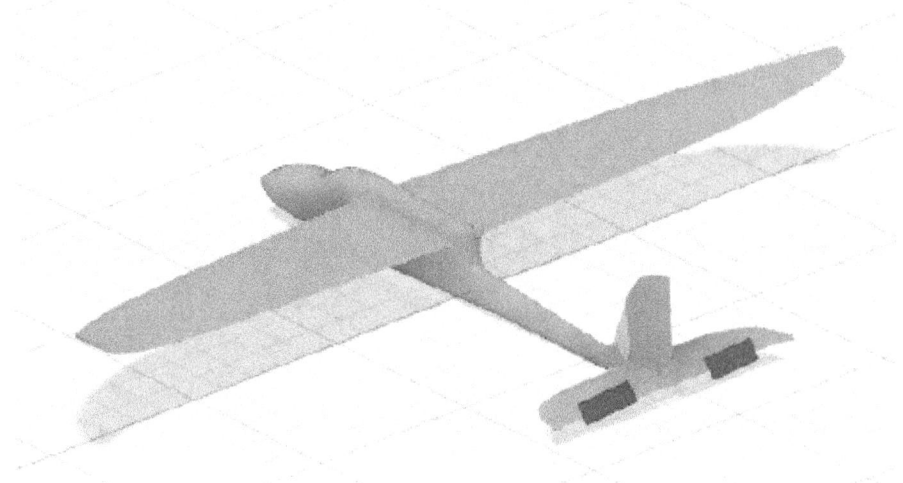

The rudder is the control surface located on the trailing edge of the vertical stabilizer (the fin) at the tail of the aircraft. It controls the aircraft's yaw (rotation around the vertical axis, which runs from the top to the bottom of the aircraft). It is operated by the rudder pedals in the cockpit. Pressing the left pedal deflects the rudder to the left, causing the nose of the aircraft to yaw to the left. Pressing the right pedal deflects the rudder to the right, causing the nose to yaw to the right. You use it to maintain coordinated flight during turns, preventing the aircraft from skidding or slipping. It also helps in counteracting adverse yaw, which is the tendency of the aircraft to yaw in the opposite direction of a roll due to differential drag on the wings.

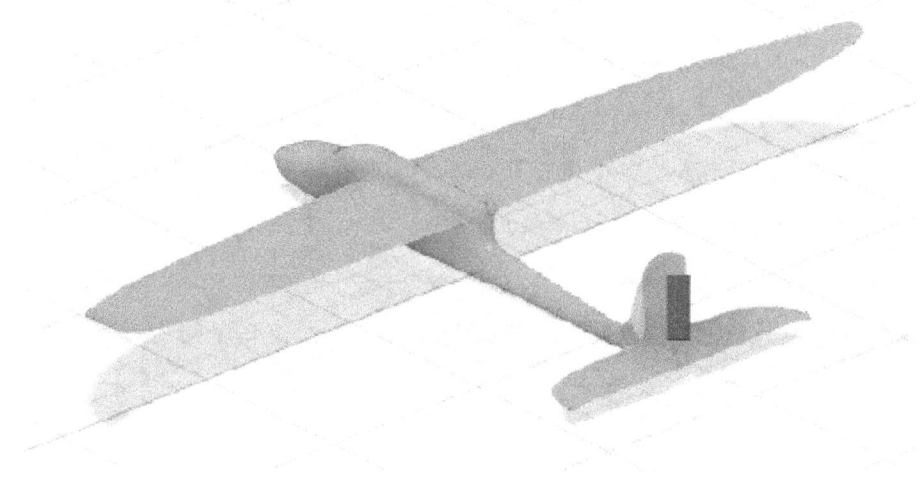

Defining Primary Control Interfaces

A quick summary here:
- The control yoke or stick manages pitch and roll by manipulating the elevators and ailerons, controlling the aircraft's nose position and banking.
- The rudder pedals control yaw by adjusting the rudder, allowing the pilot to direct the aircraft's nose left or right.
- The throttle adjusts engine power, affecting the aircraft's speed and altitude.

The control yoke, also known as the control column, or stick, as in a joystick, serves as the primary interface for managing an aircraft's pitch (the up or down movement of the nose) and roll (the banking movement to the left or right).

When it comes to pitch control, pulling the yoke or stick back raises the aircraft's nose, which causes the plane to climb, thereby increasing the pitch. Conversely, pushing the yoke or stick forward lowers the nose, resulting in a descent and a decrease in pitch. These pitch adjustments are achieved through the elevator, a control surface located on the tail's horizontal stabilizer.

For roll control, turning the yoke or moving the stick left or right tilts the aircraft's wings, causing it to roll in the respective direction, either left or right. This rolling motion is controlled by the ailerons, which are positioned on the outer trailing edges of the wings and move in opposite directions to achieve the desired bank.

The rudder pedals, positioned on the floor of the cockpit, are used to control the aircraft's yaw, which is the left or right movement of the nose around the vertical axis. To manage yaw control, pressing the left pedal moves the aircraft's nose to the left, while pressing the right pedal moves it to the right. These actions manipulate the rudder, a vertical control surface on the tail's vertical stabilizer, which adjusts the yaw and helps guide the aircraft's nose in the desired direction. Additionally, they are often used in conjunction with the yoke or stick to coordinate turns, ensuring that the aircraft maintains balance and doesn't skid or slip sideways during a turn.

The throttle is the control that manages the aircraft's engine power, directly influencing its speed and the rate at which it climbs or descends. In terms of power control, pushing the throttle forward increases the engine power, which leads to higher airspeed and potentially causes the aircraft to climb. Conversely, pulling the throttle back reduces engine power, slows the aircraft, and can cause it to descend. The throttle controls the flow of fuel and air to the engine(s), making it a crucial interface for managing the aircraft's speed and overall energy.

Defining Secondary Control Surfaces

Secondary control surfaces are additional aerodynamic components on an aircraft that assist the primary control surfaces in refining and optimizing the aircraft's performance. While the primary control surfaces (ailerons, elevator, and rudder) manage the fundamental aspects of flight such as pitch, roll, and yaw, secondary control surfaces provide finer control, improve efficiency, and enhance safety under various flight conditions. The most common secondary control surfaces are flaps, slats, spoilers, and trim tabs.

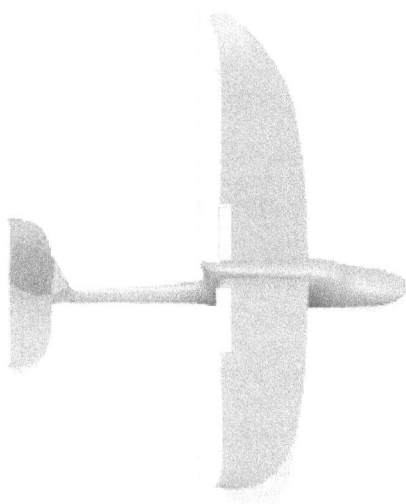

Flaps are hinged surfaces located on the trailing edge of the wings, closer to the fuselage. They are primarily used during takeoff and landing to increase the lift generated by the wing at lower speeds. They extend downward from the wing, increasing the wing's surface area and camber. This results in a higher lift coefficient, allowing the aircraft to fly safely at lower speeds, which is crucial during landing and takeoff. When deployed, they increase both lift and drag. The additional lift allows the aircraft to take off or land on shorter runways, while the increased drag helps in reducing speed during descent.

Spoilers are panels located on the upper surface of the wings, used to disrupt airflow and reduce lift. They are primarily used to assist in descent and braking. They extend upward into the airflow, creating turbulence and disrupting the smooth flow of air over the wing. This disruption reduces lift and increases drag. Note that they are typically used during descent to help the aircraft lose altitude without gaining speed. They are also deployed upon landing to "dump" lift and increase the weight on the landing gear, enhancing braking effectiveness.

Trim tabs are small adjustable surfaces attached to the trailing edges of the primary control surfaces (such as the elevator, rudder, or ailerons). They are used to maintain the aircraft in a steady state of flight without continuous input from the pilot. Simply put, they adjust the neutral position of the control surface to which they are attached. For example, if an aircraft tends to pitch up, a trim tab on the elevator can be adjusted to counteract this tendency, reducing the need for the pilot to hold the yoke or stick in a specific position constantly while making flying more comfortable.

As a quick summary:
- Elevator Trim: If the aircraft's nose consistently wants to pitch up or down, elevator trim needs to be adjusted.
- Rudder Trim: If the aircraft's nose tends to yaw left or right, rudder trim needs adjustment.
- Aileron Trim: If one wing tends to drop consistently, aileron trim needs to be adjusted.

Day 3 - Chord Line, CG, CP, AOA and Stall

Knowing the Chord Line

Knowing the chord line of an airfoil is crucial in aerodynamics because it serves as a reference point for several key measurements and concepts that are essential for understanding and controlling the performance of an aircraft. As said before, it is nothing more than an imaginary straight line drawn from the leading edge (the front) to the trailing edge (the back) of the airfoil.

The Angle of Attack (AoA) is the angle between the chord line of the airfoil and the direction of the relative airflow. This AoA is a critical factor in determining the amount of lift generated by the wing. If the AoA is too low, the wing may not generate sufficient lift; if it's too high, the airflow may separate from the wing, leading to a stall. By knowing the chord line, pilots and engineers can accurately measure and control the AoA to ensure safe and efficient flight.

The chord line helps define the orientation of the airfoil in relation to the airflow, which directly affects the aerodynamic forces acting on the wing, such as lift and drag. Understanding how these forces interact with the airfoil's orientation is essential for designing efficient wings and control surfaces.

In the design of an airfoil, the shape and length of the chord line influence the airfoil's performance characteristics, such as its lift-to-drag ratio, stall behavior, and overall efficiency. Designers use the chord line to calculate important parameters like the mean aerodynamic chord (MAC), which helps in balancing and controlling the aircraft.

The chord line can also be used to locate the center of pressure (the point where the lift force acts) and the aerodynamic center (the point along the chord line where the *pitching moment is constant). These points are vital for understanding the stability and control characteristics of the aircraft.*

The relationship between the aircraft's center of gravity (CG) and the aerodynamic center (which is referenced to the chord line) affects the aircraft's stability and control. Pilots and engineers use the chord line to assess whether the aircraft will be stable in flight and how it will respond to control inputs.

CG VS CP

The center of pressure (CP) and center of gravity (CG) are key concepts in physics and engineering, particularly in the fields of aerodynamics, structural engineering, and naval architecture. While they are often related, they represent different aspects of how forces act on a body. In aerodynamic applications, the interaction between CP and CG is crucial for stability. Designers often aim to keep the CG ahead of the CP to maintain stable flight or movement, ensuring the object (like an airplane) doesn't flip or rotate unexpectedly.

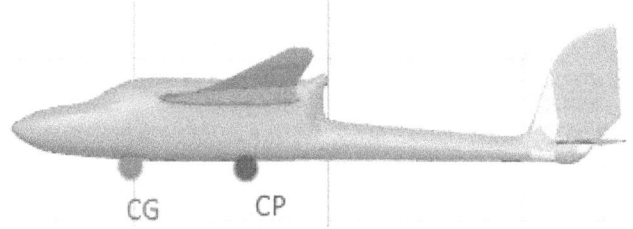

The center of pressure refers to the point where the total sum of a pressure field acts on a body, resulting in a force and possibly a moment. It represents the average location of the pressure distribution over the surface of an object and is particularly important in fluid dynamics. For example, in the design of aircraft, rockets, or ships, the center of pressure determines where aerodynamic forces like lift and drag are applied. The position of the center of pressure is not fixed; it can change depending on the shape of the object, the angle of attack, and the flow conditions of the surrounding fluid. As a result, the center of pressure is dynamic and can shift based on external conditions.

In contrast, the center of gravity is the point where the entire weight of a body is considered to act. It represents the average location of the mass distribution within an object and plays a crucial role in structural engineering, vehicle dynamics, and robotics. The center of gravity is vital for determining the stability and balance of an object. For a rigid body in a uniform gravitational field, the center of gravity is a fixed point, though its position can change if the mass distribution within the object changes, such as when cargo shifts in a vehicle.

The key difference between the two concepts lies in the nature of the forces they describe. The center of pressure is concerned with pressure forces, such as those exerted by air or water, while the center of gravity deals with gravitational forces, which are related to the mass of the object. The position of the center of pressure can move based on external pressure distribution, whereas the center of gravity is typically fixed unless the mass distribution changes.

The interaction between the center of pressure and the center of gravity is crucial for understanding the motion and stability of objects.

Center of Gravity Equals Center of Pressure

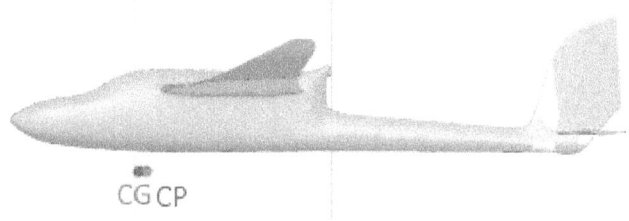

When the CG and CP coincide, the aircraft's weight acts through the same point as the aerodynamic forces. This can theoretically lead to a neutrally stable condition. The aircraft would be neutrally stable, meaning it would neither return to nor diverge from its original position after being disturbed. In practice, this condition is not ideal because even a small disturbance could make the aircraft hard to control, as there is no restoring force to return it to equilibrium.

Center of Gravity in Front of Center of Pressure

This is the most common and desired configuration for most aircraft, where the CG is ahead of the CP. It is inherently stable. If the aircraft is disturbed and the nose pitches up or down, the moment created by the CG being ahead of the CP will naturally work to return the aircraft to its original attitude. The aerodynamic forces act through the CP, and because the CG is ahead, any disturbance will create a restoring moment.

Under this configuration, the aircraft is easier to control because it will tend to self-correct when disturbed. This configuration also ensures that the aircraft is more forgiving and less prone to uncontrollable flight conditions like spins. In the real world, most aircraft are designed with the CG slightly forward of the CP to ensure stability and predictable handling characteristics.

Center of Pressure in Front of Center of Gravity

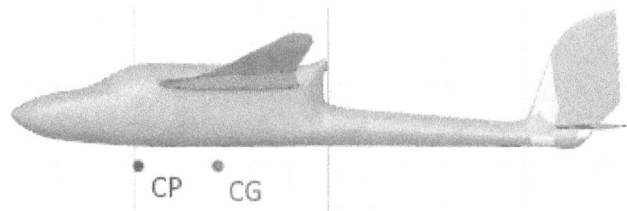

When the CP is ahead of the CG, the aircraft's aerodynamic forces act ahead of the point where the weight acts. This configuration is inherently unstable. If the aircraft is disturbed, the nose will tend to pitch further in the direction of the disturbance rather than returning to its original attitude. This is because the moment generated by the CP being ahead of the CG will amplify the disturbance, leading to a divergent situation. Controlling the aircraft in this configuration is extremely difficult and often requires active stabilization systems. The aircraft would be prone to oscillations and could easily enter uncontrollable conditions like a spin or a deep stall.

CG determination

* This is highly technical and is FYI only.

The CG represents the point where the total weight of the aircraft acts, and it has a significant impact on the aircraft's balance and handling characteristics. To determine this, the first step is to weigh the aircraft. This is typically done by placing the aircraft on a set of calibrated scales, usually three in total—one for each wheel. The aircraft must be in a level position during this process. The weight measured on each wheel, including the nose or tail wheel and the main landing gear wheels, is recorded. This data provides the weight distribution across the aircraft. Next, the arm, or the distance from a reference point on the aircraft, is measured. The reference datum is a fixed point on the aircraft from which all horizontal distances are measured. This datum is defined by the manufacturer and can be located at various points on the aircraft, such as the nose or the firewall. The horizontal distance from the datum to each wheel's contact point with the ground is measured. These measurements are known as the "arms." To calculate the moment for each wheel, the weight measured is multiplied by the corresponding arm (moment = weight × arm).

With these measurements in hand, the center of gravity is then calculated. First, the total weight of the aircraft is determined by adding up all the weights recorded from the scales. Then, the total moment is calculated by summing all the moments from each wheel. The CG is found by dividing the total moment by the total weight, which gives the distance from the reference datum to the CG.

Once the CG is calculated, it is crucial to verify it against the aircraft's CG envelope.

This envelope, defined by the manufacturer, outlines the limits within which the CG must lie for safe operation. Ensuring that the CG falls within this envelope is essential for maintaining the aircraft's stability and controllability throughout the flight. If the calculated CG is outside the allowable limits, adjustments are necessary, such as shifting cargo, adjusting fuel loads, or redistributing passengers.

It is also important to consider how the CG might change during flight, particularly as fuel is burned or payload is altered. The CG can shift, so it must be calculated for various loading conditions to ensure it remains within safe limits. This includes considering the empty weight CG and then accounting for the effects of fuel, passengers, cargo, and other variable loads to calculate the loaded CG.

For example, if an aircraft is weighed and the nose wheel registers 500 lbs, with the main gear on the left and right sides each showing 1500 lbs, and the distances from the datum to the nose wheel and main gear are 10 feet and 30 feet, respectively, the moments are calculated by multiplying the weights by their respective arms. The total weight is 3500 lbs, and the total moment is 95000 lb-ft. Dividing the total moment by the total weight gives a CG of 27.14 feet from the datum.

The Empty CG refers to the center of gravity of the aircraft when it is in its empty or basic configuration. This includes the aircraft's structure, engines, fixed equipment, and any fluids that cannot be drained (such as residual fuel and oil), and does not account for passengers, cargo, fuel (beyond residual amounts), or any other variable loads. It provides a baseline for understanding the aircraft's balance without any payload or fuel. In fact, aircraft manufacturers use the Empty CG to define the limits within which the CG must remain when the aircraft is loaded.

CG and CL

The Loaded CG determines the actual balance of the aircraft in its operational state.

It directly affects the aircraft's stability, controllability, and overall safety. Its position relative to the aircraft's center of lift CL is crucial. If the Loaded CG is too far forward or aft, the aircraft may experience stability issues, such as being nose-heavy or tail-heavy, which can make it difficult to control. CL is the single point along the aircraft's longitudinal axis (from nose to tail) where the total lift force can be considered to act. This point is essentially the average location of all the individual lift forces generated by the wings and, in some cases, the tailplane (horizontal stabilizer). It typically lies somewhere along the wing's chord line, and its exact location can shift depending on the aircraft's design, angle of attack, and flight conditions.

CL and AOA

As the angle of attack increases, the lift produced by the wing generally increases, but this also causes the center of lift to move. Typically, the center of lift moves forward with increasing angle of attack and rearward when the angle of attack decreases. Note that the position of the center of lift relative to the center of gravity is critical for the aircraft's stability. If the center of lift is ahead of the center of gravity, the aircraft can become unstable and tend to pitch nose-up, potentially leading to a stall. Conversely, if the center of lift is behind the center of gravity, the aircraft tends to be more stable, as any disturbance causing the nose to rise will be counteracted by a natural tendency to pitch down, restoring level flight.

CP and AOA

As said before, the angle of attack is the angle between the oncoming air (relative wind) and the chord line of the wing (an imaginary straight line connecting the leading and trailing edges of the wing). As the angle of attack increases, the airflow over the wing changes, increasing the lift generated by the wing. However, this also changes the pressure distribution across the wing, often moving the CP.

At low angles of attack, the CP tends to be closer to the leading edge of the wing. As the angle of attack increases, the CP typically moves rearward toward the trailing edge. At very high angles of attack (near stall conditions), the CP can move forward again, contributing to instability.

Defining Boundary Layer

The boundary layer is a thin layer of fluid, such as air or water, that forms along the surface of a solid object (like an aircraft wing or a ship's hull) as fluid flows over it. Within the boundary layer, the fluid's velocity changes from zero at the surface (due to the no-slip condition, where the fluid sticks to the surface) to the free-stream velocity of the fluid away from the surface. Note that this layer can be either laminar or turbulent. In a laminar boundary layer, the fluid flows in smooth, parallel layers with little mixing between them. In a turbulent boundary layer, the flow is chaotic and mixed, with swirls and eddies that increase friction and drag.

The behavior of the boundary layer is critical for generating lift. The shape of the airfoil is designed to control it, maintaining smooth airflow over the wing to create a pressure difference between the upper and lower surfaces. This pressure difference is what generates lift. If the boundary layer separates from the airfoil surface too early (due to an excessively high angle of attack or surface roughness), it can lead to a stall, where the lift dramatically decreases.

Flow separation occurs when the boundary layer can no longer adhere to the surface of the airfoil, typically at high angles of attack or when the boundary layer becomes too thick. This separation causes a significant increase in drag and a loss of lift, which can lead to stalling.

Flow Separation and Stall

Flow separation is the precursor to a stall. Flow separation occurs when the smooth, laminar airflow over the surface of a wing or any other aerodynamic surface becomes disturbed and detaches from the surface. Normally, as air flows over the wing, it closely follows the wing's contour. The wing's shape, or airfoil, is specifically designed to create a pressure difference—lower pressure above the wing and higher pressure below—which generates lift. Under ideal conditions, this airflow remains attached to the wing's surface, flowing smoothly from the leading edge (the front of the wing) to the trailing edge (the back).

However, when the angle of attack—the angle between the wing's chord line and the oncoming airflow—becomes too steep, the airflow can no longer follow the wing's contour smoothly. At this point, the airflow begins to separate from the wing's surface. This separation creates turbulent airflow behind the separation point, forming a chaotic region known as a wake.

A stall is a consequence of extensive flow separation over a significant portion of the wing, leading to a dramatic loss of lift. As the angle of attack increases, the point where flow separation begins moves forward along the wing. Once enough airflow has separated, the wing can no longer generate sufficient lift to support the aircraft's weight. A stall occurs when the lift generated by the wing suddenly drops, causing the aircraft to lose altitude. This happens because the smooth airflow, essential for generating lift, has broken down across a large part of the wing. Stalls are typically associated with high angles of attack but can also occur at low airspeeds or if the wing's surface is disrupted by ice or other contaminants.

Day 4 - More on Lift & Drag, Wingtip vortices, Downwash, Ground Effects and Landing

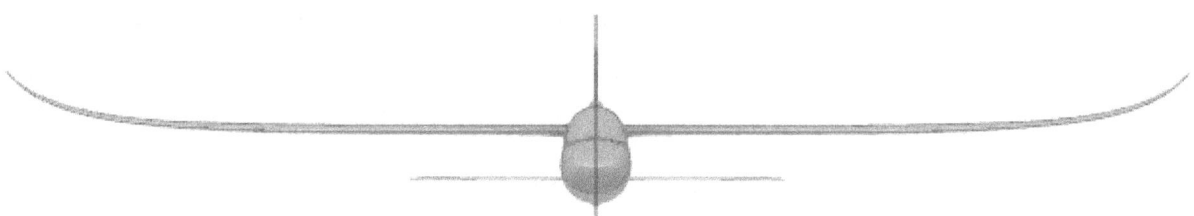

Aerodynamic center

The aerodynamic center is a critical point on an aircraft's airfoil (such as a wing) where the pitching moment (the tendency of the airfoil to rotate about its center of gravity) remains constant regardless of changes in the angle of attack. In simpler terms, it's the point along the chord line of an airfoil where the aerodynamic forces effectively act, and the moment about this point does not change with varying lift.
Key Concepts:

For most airfoils, the aerodynamic center is typically located near the quarter-chord point, or 25% of the distance from the leading edge of the airfoil. This is true for many conventional subsonic airfoils. Note that the location of the aerodynamic center relative to the center of gravity (CG) is crucial for stability. If the aerodynamic center is behind the CG, the aircraft tends to be more stable, as the moment arm creates a nose-down moment that helps return the aircraft to equilibrium after a disturbance.

Relative Wind and AOA VERSUS Critical AoA

Relative wind refers to the airflow that is opposite in direction to the flight path of an aircraft or the motion of an airfoil (such as a wing or propeller blade). It is the direction of the air that the aircraft experiences as it moves through the atmosphere.

It is directly opposite to the direction of flight - in fact the AOA is measured relative to this wind direction.

We need to stress that AoA is the angle between the chord line of the wing (again, this is an imaginary straight line from the leading edge to the trailing edge) and the relative wind. This value is a runtime dynamic parameter. The critical AOA, however, is a design value which is fixed.

The critical AoA is the maximum angle of attack at which an aircraft's wing can produce lift (you should not attempt to exceed it). Beyond this angle, the airflow over the wing separates, causing a dramatic loss of lift and leading to a stall. This is a specific angle for a given wing design and does not change with speed, weight, or other factors. It typically ranges between 15° to 20° for most aircraft wings. When the wing exceeds this, it enters a stall regardless of the airspeed or power setting. This is because the smooth, laminar airflow over the wing is disrupted, and lift can no longer be effectively generated.

Negative AOA and why

A negative AoA often occurs in descending flight when the nose of the aircraft is pointed downward relative to the oncoming airflow. During certain maneuvers or when descending, the wing may have a negative AoA, where the airflow hits the bottom surface of the wing first, which can reduce lift or could even produce negative lift (downward force).

Note that while stalls typically occur at positive AoA when the airfoil exceeds its critical AoA, it's also possible, though less common, for a stall to occur at a negative AoA if the airflow over the wing becomes sufficiently disrupted.

Airspeed and AoA

At lower airspeeds, the aircraft requires a higher AoA to generate enough lift to counteract its weight. This is because lift is proportional to the square of the airspeed (Lift Airspeed^2). At slower speeds, the lift produced by the airflow over the wing is less, so the AoA must be increased to generate the required lift. Do beware, while increasing AoA allows the wing to generate more lift up to the critical AoA, beyond which the airflow will separate from the wing surface, leading to a stall. Managing AoA is therefore very crucial to avoid stalling at low speeds.

At higher airspeeds, the airflow over the wing is faster, increasing the lift produced without needing to increase AoA. Because lift increases with the square of the airspeed, higher speeds generate more lift even at a lower AoA. With higher airspeed, the wing can generate the necessary lift with a smaller AoA, which is more aerodynamically efficient and produces less drag compared to a higher AoA.

AOI Angle of Incidence

The angle of incidence is the angle between the chord line of an aircraft's wing (or other airfoil) and the longitudinal axis of the aircraft. The longitudinal axis runs along the length of the fuselage, from the nose to the tail. It is a fixed structural angle, meaning it is set during the aircraft's design and construction and cannot be adjusted by the pilot during flight. In real world it is usually small, often between 0 and 5 degrees.

Wing surface area and its relationship with drag & lift

The wing surface area is the total area of the wing that is exposed to the airflow. It is a critical factor in determining the lift produced by the wing. Larger wing surface areas generally produce more lift because more air is displaced by the wing. This means that for a given AoA, a wing with a larger surface area will generate more lift than a smaller wing.

Aircraft with larger wings can generate more lift at lower speeds and lower AoA, which is useful for maintaining flight efficiency and reducing the risk of stall. Smaller wings may need higher speeds or higher AoA to generate the same amount of lift. HOWEVER, a larger wing surface area means there is more surface for the air to move over, which increases the skin friction drag. This is because a larger area leads to more contact between the air and the wing, generating more resistance. The overall shape of the wing creates drag as the wing pushes through the air. A larger surface area generally means a bigger cross-sectional area, which can increase form drag. Larger wings can also lead to higher induced drag. This is because induced drag is related to the generation of lift, and while larger wings can produce more lift, they also increase the downward deflection of air (vortex drag) at the wingtips. Induced drag is more significant at lower speeds and higher angles of attack, where the wing generates more lift relative to its size.

Simply put, the relative wind does not "push" the wing backward in a literal sense. Instead, it serves as a reference for the direction in which aerodynamic forces act. In fact, drag (being the force that resists the forward motion) acts in the direction opposite to the aircraft's motion, which can somehow be interpreted as pushing the wing backward in terms of slowing the aircraft down.

Wingtip vortices, downwash, ground effects, and landing

Wingtip vortices, downwash, ground effects, and landing are all interrelated concepts in aviation that involve the interaction between an aircraft's wings and the surrounding air, especially during the critical phases of flight like takeoff and landing.

Wingtip vortices are spiraling air patterns that form at the tips of an airplane's wings. When a wing generates lift, the high-pressure air beneath the wing flows outward and around the wingtips to the low-pressure region above the wing. This movement creates a circular motion, or vortex, at the wingtips, with the air spiraling outward from the bottom and inward on top. Wingtip vortices are a natural consequence of lift generation but they create induced drag, a form of aerodynamic drag that opposes the aircraft's forward motion, reducing overall efficiency.

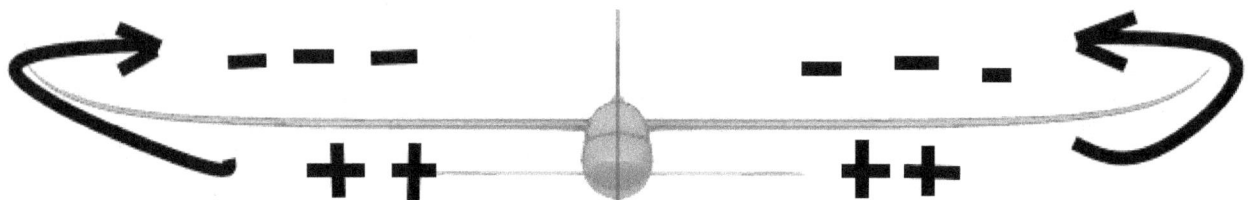

The presence of wingtip vortices also contributes to downwash, which is the downward deflection of air as it passes over the wing. To be precise, downwash is the downward deflection of airflow as it passes over and beyond the wings of an aircraft, and it is a direct consequence of how wings generate lift. As the wing generates lift, it pushes air downward, and this downwash is most noticeable directly behind the trailing edge of the wing. Downwash affects the effective angle of attack of the wing—the angle between the chord line of the wing and the relative airflow—by slightly reducing it, which in turn affects the amount of lift and drag produced by the wing. The interaction of wingtip vortices and downwash is particularly important because it influences how efficiently the aircraft flies, with increased downwash leading to increased induced drag.

In the broadest sense, downwash is neither inherently good nor bad—it is simply a byproduct of how wings create lift. Without downwash, there would be no lift, and the aircraft would not be able to fly. However, downwash also contributes to induced drag, which is generally considered an undesirable effect.

When an aircraft is close to the ground, typically within one wingspan of the surface, it experiences what is known as ground effect. Ground effect occurs because the proximity of the ground interferes with the formation of wingtip vortices and reduces the strength of the downwash. This interference leads to a reduction in induced drag and an increase in lift, which can make the aircraft feel more buoyant and stable as it approaches the runway. Ground effect is most pronounced during landing and takeoff. Pilots must account for ground effect because it can cause the aircraft to float or stay airborne longer than expected, making landing more challenging as it reduces the rate of descent.

During landing, as the aircraft descends towards the runway, it enters ground effect. The reduction in drag and increase in lift can make it more difficult to lose altitude, requiring precise control to ensure a smooth touchdown. Pilots often experience a sensation of the aircraft "floating" just above the runway as ground effect temporarily boosts lift. Managing this effect is critical to executing a safe and controlled landing. To counteract the increased lift and reduce the aircraft's tendency to float, pilots may adjust the throttle to decrease speed and ensure the aircraft settles onto the runway smoothly.

Is downwash good or bad?

Downwash is the downward deflection of air behind the wing, which occurs as a natural consequence of the wing's interaction with the air. It normally reduces the effective angle of attack and thus the lift. Simply put, it is essential for creating lift, but it also leads to induced drag, which is a negative effect. The goal in aerodynamics is often to manage downwash to maximize lift while minimizing induced drag.

Ground effect during landing

As previously mentioned, ground effect is a phenomenon that occurs when an aircraft is flying close to the ground, typically at an altitude equal to or less than the wingspan of the aircraft. This effect results in a noticeable increase in lift and a decrease in drag, making it particularly relevant during landing and takeoff.

The high-pressure air below the wing is less able to "spill" around the wingtips, resulting in more effective lift production - in other words, air below the aircraft becomes compressed. This is why the aircraft experiences an increase in lift as it descends into ground effect. Also, when near the ground the downwash is less pronounced because the ground disrupts the airflow pattern. This blocks the air from correcting the pressure differential.

As the aircraft enters ground effect during landing, the sudden increase in lift can cause the aircraft to "float" above the runway, making it harder to bring the aircraft down for a smooth touchdown.

Ground effect during takeoff

Ground effect can shorten the takeoff distance required for the aircraft to achieve lift-off. This can be particularly beneficial in scenarios where runway length is limited. Pilots often experience a "floating" sensation as the aircraft approaches the ground effect zone. This is because the increase in lift makes the aircraft feel as if it is lifting off the ground sooner than expected.

As the aircraft climbs out of ground effect, the lift boost and drag reduction disappear. The pilot needs to manage the transition smoothly to avoid an abrupt loss of lift or an excessive rate of climb. Once out of ground effect, the aircraft will need a higher angle of attack to maintain climb performance. Pilots must be prepared for this transition to ensure a smooth climb out.

Day 5 - Stall Speed, Prop Bias, Stabilizers and Flight Stability

Stall speed

Stall speed is the minimum speed at which an aircraft can maintain level flight before it stalls (you should not go slower than this). At this speed, the aircraft's wings can no longer generate sufficient lift to counteract its weight, leading to a loss of controlled flight. Generally, heavier aircraft have a higher stall speed because more lift is required to support the weight. As an aircraft banks, the stall speed increases because more lift is required to maintain altitude. Also, during maneuvers, the load factor increases, requiring more lift and thus increasing stall speed.

Do note that the stall speed is usually indicated on the aircraft's airspeed indicator as a specific value (Vs). However, this is indicated airspeed and can differ from the true airspeed due to factors like altitude and air density.

Stall and spin

A spin typically begins with a stall, which occurs when the AoA exceeds the critical angle, causing the airflow over the wing to separate and lift to decrease dramatically. In a spin, the stall is not symmetric. One wing stalls more severely than the other. The wing that is more stalled produces less lift and more drag, while the other wing produces more lift and less drag. This imbalance causes the aircraft to yaw and roll in the direction of the stalled wing. As the aircraft yaws and rolls, it begins to descend in a spiraling path, with the nose pointing down and the aircraft rotating around its vertical axis. A spin can occur with the engine at full power (power-on spin) or with the engine at idle (power-off spin). The dynamics of the spin can differ depending on the power setting, with a power-on spin often being more nose-high and slower than a power-off spin.

Propeller plane VS Jet stream plane

Propeller planes and jet planes differ primarily in their propulsion systems, which impacts their performance and typical use.

A propeller plane generates thrust through one or more propellers, which are powered by engines—either piston engines or turboprops. The propeller spins to move air backward, creating the force that propels the aircraft forward. Propeller planes typically fly at lower speeds, usually ranging between 100 to 300 knots (115 to 345 mph), and they operate at lower altitudes, often between 10,000 to 25,000 feet. These planes are more fuel-efficient at these lower speeds and altitudes, making them ideal for short to medium-range flights, regional transportation, and general aviation. Common examples of propeller planes include the Cessna 172, Piper PA-28, and the de Havilland Canada Dash 8.

On the other hand, jet planes are powered by jet engines, which work by compressing air, mixing it with fuel, and igniting the mixture. This combustion process produces a high-speed exhaust that generates the thrust needed to propel the plane forward. Jet planes are capable of much higher speeds than propeller planes, typically cruising between 400 to 600 knots (460 to 690 mph) or even faster. They also operate at higher altitudes, generally between 30,000 to 40,000 feet, where the thinner air reduces drag and improves fuel efficiency. Jet planes are designed for long-haul flights, fast travel, and high-altitude operations. Examples of jet-powered aircraft include the Boeing 737, Airbus A320, and the F-16 Fighting Falcon.

Prop Wash

Prop wash, also known as slipstream, is the turbulent airflow that is generated behind a propeller when it is spinning. This airflow is typically directed backward along the fuselage and wings (thus influencing various aerodynamic surfaces) and is the result of the propeller accelerating the air backward as it provides thrust to move the aircraft forward. It can affect nearby aircraft on the ground, especially during takeoff or taxiing. It can also impact the control surfaces of the aircraft producing it, particularly the tail surfaces, which might be subjected to increased airflow and turbulence.

Propeller plane left turning tendency

There is a tendency of the propeller aircraft to yaw, or rotate, to the left during certain phases of flight, particularly during takeoff or when the engine is under high power. This phenomenon is primarily due to a combination of aerodynamic and mechanical forces acting on the plane, specifically in propeller-driven aircraft.

- P-Factor (Asymmetric Thrust): When a propeller plane is in a nose-up attitude (such as during takeoff), the downward-moving blade on the right side of the propeller disk generates more thrust than the upward-moving blade on the left side. This asymmetry causes the aircraft to yaw to the left.

- Torque Effect: As the propeller spins clockwise (when viewed from the cockpit in most planes), it generates an equal and opposite reaction that causes the aircraft to roll to the left. This roll can contribute to a leftward yaw, particularly in single-engine planes.
- Spiraling Slipstream: The propeller generates a spiraling slipstream of air that wraps around the fuselage and strikes the left side of the aircraft's tail. This force pushes the tail to the right, causing the nose of the plane to yaw to the left.
- Gyroscopic Precession: When the aircraft's pitch changes (like during takeoff or climbing), the rotating propeller acts as a gyroscope. The precession effect of the gyroscope causes a force that tends to yaw the aircraft to the left.

P-Factor (Asymmetric Blade Effect)

P-Factor, or asymmetric blade effect, refers to the phenomenon where the descending blade of a propeller produces more thrust than the ascending blade during forward flight, causing a yawing moment. This effect is more pronounced in high-power, low-speed situations, such as during takeoff or climb.

When an aircraft is flying at an angle of attack (nose-up attitude), the downward-moving blade on the right side (in a clockwise rotating propeller) has a higher angle of attack relative to the oncoming air than the upward-moving blade on the left. Due to the higher angle of attack, the descending blade generates more lift (thrust) than the ascending blade. This imbalance in thrust between the blades creates a yawing force that pushes the nose of the aircraft to the left (for a clockwise rotating propeller), requiring right rudder input to counteract.

Therefore, this left turning tendency really depends on the direction of prop rotation. Most single-engine aircraft, especially those produced in the United States and many other countries, have propellers that rotate clockwise when viewed from the cockpit (pilot's perspective). This is the standard direction for propeller rotation on many aircraft, including popular models like the Cessna 172, Piper Cherokee, and many others.

Considering all these together...

Pilots need to be aware of these forces and compensate for them, especially during critical phases of flight like takeoff. This is usually done by applying right rudder input to counteract the left-turning tendency. The amount of compensation required can vary based on the power setting, aircraft speed, and attitude.

Modern propeller designs sometimes incorporate blades with a slight twist or varying pitch angles that help distribute thrust more evenly and reduce the asymmetry that contributes to left bias. The engine itself is often mounted at a slight angle to the right (typically a few degrees). This small offset helps balance the asymmetric forces produced by the propeller, particularly the P-factor, by generating a compensatory force that reduces the left yaw.

Many aircraft have a slightly offset vertical stabilizer (fin) or rudder. This offset is designed to counteract the leftward yaw caused by the propeller's forces. By being slightly angled to the right, the rudder helps neutralize the left-turning tendency, especially during high-power settings like takeoff.

Some aircraft are equipped with adjustable rudder trim tabs, small surfaces on the trailing edge of the rudder that can be set to counteract the left yaw. Pilots can adjust these tabs in-flight to maintain straight and level flight without constantly applying pressure to the rudder pedals. On twin-engine planes, designers sometimes use counter-rotating propellers, where the propellers on each engine spin in opposite directions. This design cancels out the left-turning tendencies of each engine, as the forces generated by one propeller counteract those of the other.

Load factor

The load factor of an aircraft is a measure of the amount of force exerted on the aircraft's structure relative to its weight. It is typically expressed as a multiple of the force of gravity (g). The load factor indicates how much more (or less) force is being exerted on the aircraft compared to the normal force of gravity when the aircraft is in straight-and-level flight.

In straight-and-level flight, the load factor is 1g, meaning the lift generated by the wings is equal to the aircraft's weight. The force exerted on the aircraft and its occupants is equal to their normal weight.

During maneuvers such as turns, climbs, or turbulence, the load factor increases. For example, in a level, steep turn, the wings must generate more lift to maintain altitude, increasing the load factor. A load factor of 2g means the force on the aircraft is twice its weight. HOWEVER, in some maneuvers such as descending or in a push-over (nose-down) maneuver, the load factor can be less than 1g, causing a sensation of reduced weight (sometimes referred to as "negative G" if it goes below zero).

Note that aircraft are designed with specific load factor limits, beyond which structural damage or failure may occur. High load factors can also cause discomfort to passengers.

Vertical Stabilizer

The vertical stabilizer, often referred to as the fin, is a crucial component of an aircraft's tail section, playing a vital role in maintaining stable and controlled flight. Its primary function is to provide directional stability, ensuring that the aircraft remains aligned with its flight path and does not unintentionally yaw (rotate around its vertical axis) left or right. This stability is essential for keeping the aircraft flying straight and preventing unwanted deviations in direction.

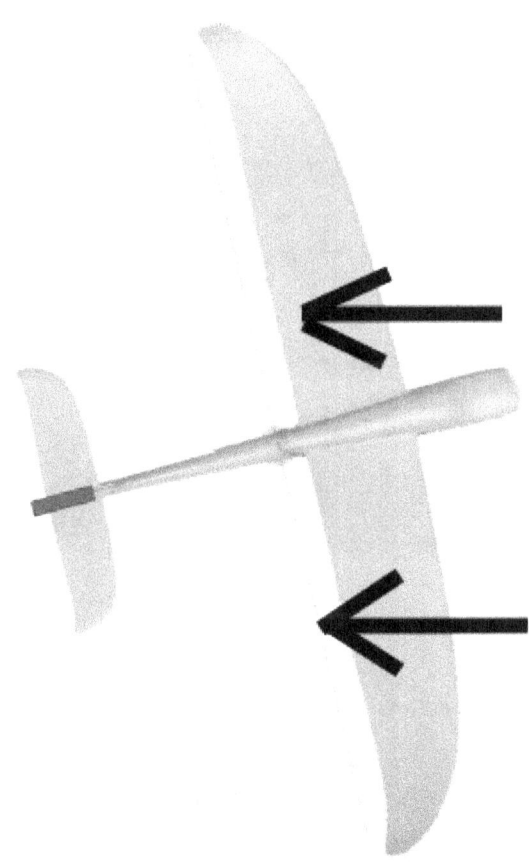

The vertical stabilizer achieves this by generating aerodynamic forces that resist any sideways movement of the aircraft's nose. When the aircraft begins to yaw, the vertical stabilizer, which is positioned vertically at the rear of the aircraft, interacts with the oncoming airflow. This interaction produces a force that counteracts the yaw, pushing the tail back in line with the direction of flight. In this way, the vertical stabilizer keeps the aircraft's nose pointed forward, helping to maintain the desired flight path.

Additionally, the vertical stabilizer typically houses the rudder, a movable control surface that pilots use to intentionally control yaw. By deflecting the rudder to the left or right, the pilot can steer the aircraft in the desired direction. The vertical stabilizer thus not only provides passive stability but also enables active directional control, allowing the pilot to make precise adjustments to the aircraft's heading.

Horizontal Stabilizer

The horizontal stabilizer is a key component of an aircraft's tail section, responsible for maintaining stability and control in the pitch axis, which is the up-and-down movement of the aircraft's nose. This stabilizer is positioned horizontally at the rear of the aircraft and plays a critical role in ensuring that the aircraft flies smoothly and remains level.

The primary function of the horizontal stabilizer is to provide longitudinal stability, meaning it helps keep the aircraft's nose from pitching up or down unexpectedly. During flight, various forces act on the aircraft, including lift generated by the wings and changes in weight distribution. Without the horizontal stabilizer, these forces could cause the nose of the plane to pitch excessively, leading to an unstable flight path.

The horizontal stabilizer can be thought of as an "upside-down airfoil" compared to the main wings of the aircraft. Simply put, it counters these forces by generating a downward aerodynamic force at the tail, which balances the lift generated by the wings. This balance ensures that the aircraft maintains a level attitude, preventing the nose from rising or falling unintentionally. By keeping the aircraft's nose aligned with the intended flight path, the horizontal stabilizer helps maintain a smooth and steady flight.

In addition to providing stability, the horizontal stabilizer also includes a movable control surface known as the elevator. The elevator allows the pilot to actively control the pitch of the aircraft. By moving the elevator up or down, the pilot can change the angle of the horizontal stabilizer, which in turn raises or lowers the nose of the aircraft. This control is essential during various phases of flight, such as takeoff, climbing, cruising, descending, and landing.

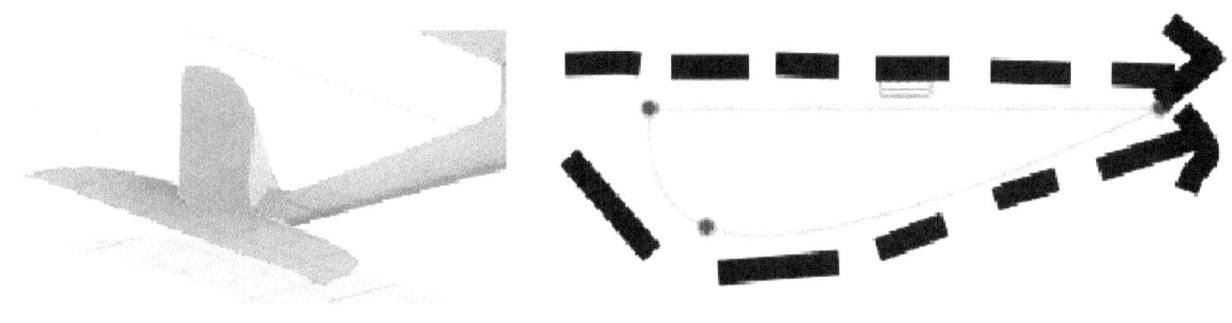

END OF BOOK

Please email your questions and comments to admin@Tomorrowskills.com.